M9 ACE

Armored Combat Earthmover

In Detail

The history, development and operations of the United States Army's armored combat earthmover engineering vehicle. Includes 132 pages and several hundred photographs.

SABOT Publications

The M9 ACE in Detail
SP002

Sabot Publications
941 Cottontail Lane
St. Charles, Missouri 63303
United States of America
ISBN: 978-0-9973774-1-5

Special thanks to our friends for their valuable contributions to this book:
CSM Mike Moske (ret)
SFC Gwynn
SSG Jackson
And the rest of the TDC 1st Brigade 35th Engineers
Charlie Pritchett

Combat engineers everywhere
and all of the Soldiers, Sailors, Marines and Airmen who serve
Thank you!

Chris Mrosko
Brett Avants
SABOT Publications LLC

Printed in the United States of America

Introduction

The M9 ACE is a highly mobile, armored tracked earthmoving vehicle that provides combat engineer support to front-line Army and Marine forces. Its capabilities include eliminating enemy obstacles, maintaining and repairing roads and supply routes and construction of fighting positions. It accomplishes several critical battlefield engineering tasks such as preparing hull defilade fighting positions for guns and tanks, and preparing protected positions for other critical battlefield systems to increase their survivability against a variety of threats.

The M9 can also repair roads, create roads for supply routes, remove roadblocks and other obstacles, and prepare access routes at water obstacles. It also performs countermobility tasks such as preparing antitank ditches and destroying/interdicting roadway networks. The M9 ACE has a crew of one and can reach a speed of 30 mph on land. The system is configured to be highly mobile, and can be airlifted in C130, C17 and C5 aircraft.

The M9 can be used as an earthmover in a variety of mission configurations. These include bulldozing, rough grading, excavating, hauling and scraping. The M9 can also be used as a cargo carrier and a prime mover. It has a drawbar pull of 31,000 lbs (14,074 kg) at 1.5 mph (2.4 km/h).

The engine, drive train and driver's compartment are in the rear of the vehicle. The front of the M9 contains an 8.7 cubic yard (6.7 m^3) box, apron and dozer blade. Its armor consists of welded aluminum with selected steel and laminated plates, which will protect the operator and vehicle against small arms and artillery fragmentation. An armored cupola containing eight vision blocks sits atop the driver's compartment. The vehicle hull is a welded and bolted aluminum structure with a two-speed winch capable of 25,000 pound (110 kN) line pull.

The M9 has a unique suspension system that allows the front of the vehicle to be raised, lowered, or tilted to permit various dozing, excavating, rough grading and ditching functions. The main components of this system are eight high-pressure hydraulic rotary actuators (four on each side) that connect to the roadwheels. During on and off-road travel, this system provides the operator with a smooth ride through the use of shock-absorbers. In earthmoving operations, the operator rotates the actuators, thus lowering the apron and blade for digging.

In addition, the M9 has a smoke screening capability in the form of smoke grenade launchers and chemical-biological protection for the operator.

A typical combat engineer battalion will contain 22 ACEs, with seven per company plus an operational readiness float. At last count, the U.S. Army had a total of 447 M9 ACEs, plus just over 100 in use with the United States Marine Corps.

The M9 ACE was also plagued with several mechanical and tactical issues originating back to the 1980's. Recently, however, the Army and the Marine Corps have been upgrading and improving the M9. The vehicle's recent improvements include an entire redesign of the hydraulic subsystem, incorporation of an automated track tensioning system, a redesign of the electrical subsystem, design and integration of a vision system consisting of ten cameras, the addition of an automated fire extinguishing system, and incorporating improvements from the Army's Standard Improvement Package-4 (SIP-4).

The ACE's hydraulic system is much more capable of operating without failure. The integrated vision system adds a new level of improved visibility for the single operator. There are cameras and lights that surround the ACE and the infrared cameras allow the operator to see through clouds of dirt and dust, including the ability to see at night. The newer cameras link to screens inside the operator's cab for enhanced visibility, which brings the visibility factor from about 180 degrees, with the hatch closed, to approximately 360 degrees.

The M9 ACE was first fielded in 1986 and saw its first combat in Operation Desert Storm in 1991, and has been used in numerous areas of operations including Iraq and Afghanistan.

From the M9 ACE Technical Manual

LOCATION AND DESCRIPTION OF MAJOR COMPONENTS

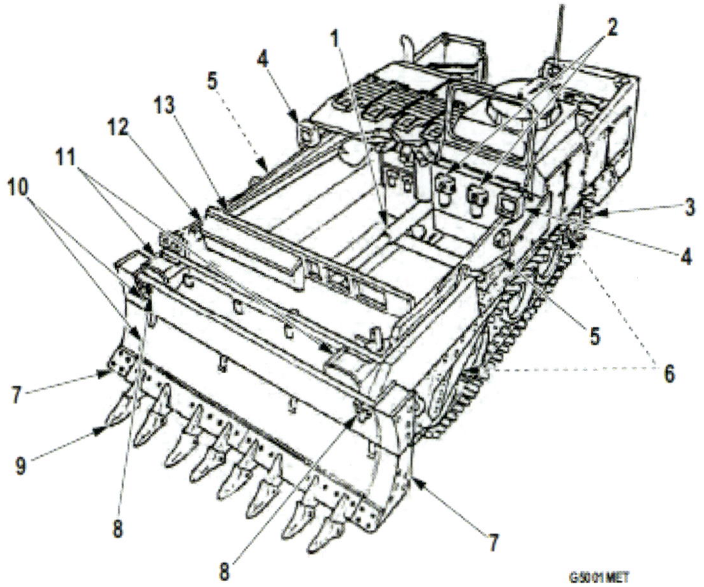

Figure 1. Location and Description of Major Components.

1. DEBRIS SHIELD (Figure 1, Item 1). Prevents debris from accumulating on hose assemblies and fittings in lower bowl area.

2. HIGH-PRESSURE FILTERS (Figure 1, Item 2). Provide increased filtration and ease of maintenance with pop-out indicators, spin-off filters, and replaceable filter elements.

3. TRACK (Figure 1, Item 3). Consists of rubber-padded steel track shoes and is driven by the sprockets.

4. FLOODLIGHTS (Figure 1, Item 4). Provide light for night operations.

5. APRON CYLINDER (Figure 1, Item 5). Raises and lowers apron and dozer assembly.

6. TRACK WEAR PLATES (Figure 1, Item 6). Replaceable steel plates keep track from wearing welded parts on the aluminum hull.

7. APRON AND DOZER EXTENSIONS (Figure 1, Item 7). Provide a larger working surface for apron and dozer. May be removed for transporting vehicle.

8. DOZER/RIPPER BLADE LATCHES (Figure 1, Item 8). Reusable pin assembly locks dozer/ripper blade in folded position during long motor marches and cross-country driving.

9. RIPPER BLADE (Figure 1, Item 9). Provides ease of digging in hard-packed soils (if installed).

10. APRON AND DOZER ASSEMBLY (Figure 1, Item 10). Used for earthmoving operations. Can be raised, or lowered to load or unload material or cargo.

11. HEADLIGHTS (Figure 1, Item 11). Provide light for night driving. Include blackout and infrared lights.

LOCATION AND DESCRIPTION OF MAJOR COMPONENTS - Continued

12. STOWAGE BOX (Figure 1, Item 12). Used to stow Basic Issue Items (BII) and Additional Authorization List (AAL) items.

13. EJECTOR (Figure 1, Item 13). Moves back and forth in bowl to discharge material, load or unload palletized cargo, and fold the dozer blade.

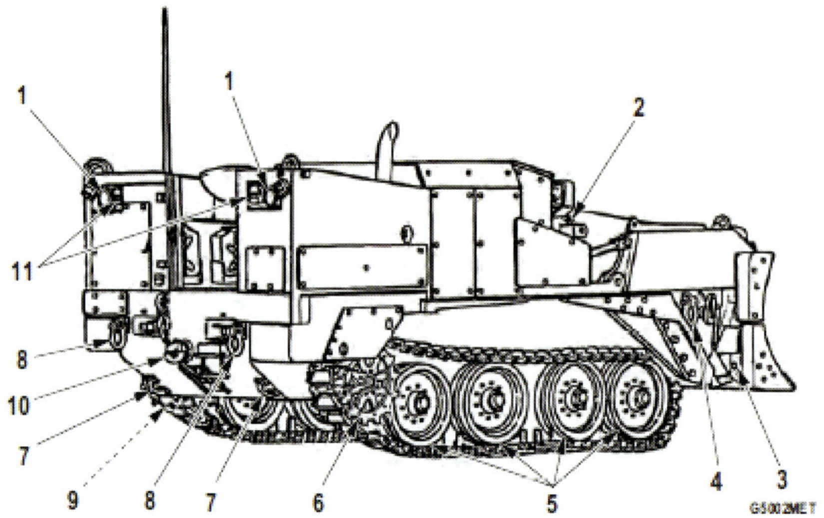

Figure 2. Location and Description of Major Components.

1. REAR FLOODLIGHTS (Figure 2, Item 1). Provide light at rear of vehicle for night operations.

2. UPPER APRON LOCKPINS (Figure 2, Item 2). Lock apron in raised position as a safety precaution during maintenance.

3. DOZER LOCKPINS (Figure 2, Item 3). Secure dozer/ripper blade to apron.

4. TIEDOWN SHACKLES (Figure 2, Item 4). Provided to secure the vehicle during air and surface transportation.

5. ROADWHEELS (Figure 2, Item 5). Provide support and guide for the tracks and suspension of the vehicle.

6. SPROCKETS (Figure 2, Item 6). Mounted to the final drives to drive the track.

7. TIEDOWN BRACKETS (Figure 2, Item 7). Provided to secure the vehicle during air and surface transportation.

8. TOW SHACKLES (Figure 2, Item 8). Used to attach tow cable, or chain, for towing or recovery operations. Also used to secure the vehicle during surface transportation.

9. HULL PROTECTIVE PLATES (Figure 2, Item 9). Protect hull underside from damage during digging operations.

10. TOWING PINTLE (Figure 2, Item 10). Used during towing operations to attach tow bar or to tow trailer.

11. TAILLIGHTS (Figure 2, Item 11). Used for night driving and to indicate when vehicle brakes are used.

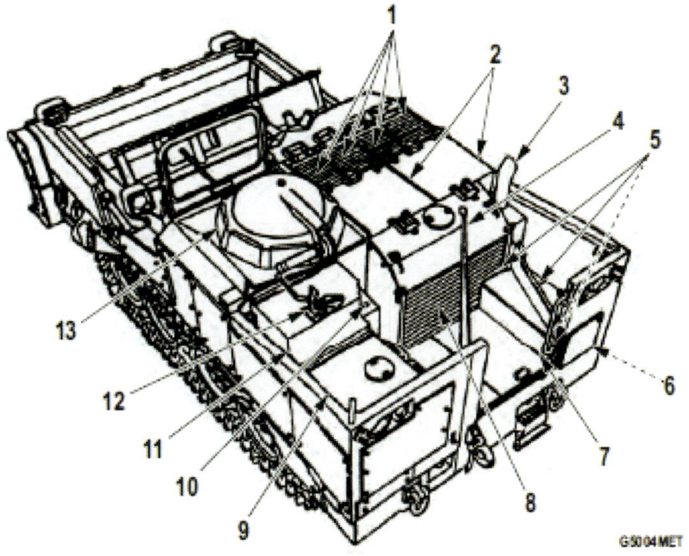

Figure 4. Location and Description of Major Components.

1. INTAKE GRILLES (Figure 4, Item 1). Louvered armor plates provide passage of air for engine cooling while protecting engine and engine components from projectiles.
2. ACCESS COVERS (Figure 4, Item 2). Non-louvered armor plates to protect engine.
3. MUFFLER (Figure 4, Item 3). Reduces noise level of vehicle exhaust system.
4. ANTENNA (Figure 4, Item 4). Allows transmission and reception for the vehicle radio.
5. BATTERY BOX DEFLECTORS (Figure 4, Item 5). Deflect radiator hot air flow away from battery box, preventing boil-over in high ambient temperatures.
6. SLAVE RECEPTACLE (Figure 4, Item 6). Connector for slave starting the vehicle from an external source.
7. BATTERY BOX (Figure 4, Item 7). Protects the four vehicle batteries.
8. EXHAUST GRILLES (Figure 4, Item 8). Louvered armor plates provide passage of air for engine cooling while protecting radiator from projectiles.
9. FUEL TANK (Figure 4, Item 9). Contains fuel for vehicle engine operation.
10. HYDRAULIC OIL FILL PORT (Figure 4, Item 10). Primary fill point for servicing the vehicle hydraulic tank.
11. RADIO BOX (Figure 4, Item 11). Provides protection for the radio.
12. HATCH COVER LATCH (Figure 4, Item 12). Used during noncombat situations to hold down open hatch cover to prevent accidental closing of hatch cover.
13. OPERATOR'S HATCH ASSEMBLY (Figure 4, Item 13). Armored cover for operator's compartment. Contains eight vision blocks and can be latched open or locked closed.

SLING
TIE DOWN
B402
TIE DOWN

The M9 ACE *In Action*

DIRT DIGGLER

Two armored combat excavators from the 43rd Combat Engineer Company, 3rd Squadron, 3rd Armored Cavalry Regiment, work as a team to remove debris from Highway 1 during Operation Thunder Reaper which has cleared 15 Kilometers of road for public use in just over 30 days, Mosul, Iraq, Dec. 22.

An Armored Combat Excavator from 43rd Combat Engineer Company, 3rd Armored Cavalry Regiment, clears debris from a major thoroughfare in Mosul, Iraq, during Operation Thunder Reaper, Dec 18. Operation Thunder Reaper, a route clearance operation, has cleared and repaired 5 kilometers of road for use by the citizens of Mosul.

Above: The 309th Engineer Company (Mobility Augmentation Company) out of Brainerd, Minn., brings an M9 Army Combat Earth Mover brings a Mine Clearing Line Charge up for test firing at Fort McCoy, Wis., July 22. The engineers are at Fort McCoy taking part in Warrior Exercise 86 for their Extended Combat Training, which gives them the opportunity to practice their Tactics, Techniques and Procedures. (U.S. Army photo by Sgt. Francis Horton, 363rd Public Affairs Detachment)

Sgt. Robert Kubach, the operations chief for 2nd Combat Engineering Battalion and Cleveland native, operates the improved M9 Armored Combat Earth-mover during a familiarization exercise Feb. 26, 2014. The M9 ACE is the improved version of the Legacy ACE; it has an upgraded and reinforced hull, more powerful engine and improved hydraulic system.

Left: Eighth Engineer Support Battalion's Marine Sgt. Anthony Queisser (right) discusses the M9 Armored Combat Earthmover's (ACE) automatic track tension system with Heavy Equipment Platoon Commander Chief Warrant Officer Brad Pottorff at Aberdeen Test Center, Aberdeen, Md. Right: The M9 ACE pictured above is one of the six low rate initial production vehicles used in testing by the Marines. (Photo courtesy of U.S. Army)

Above and below: U.S. Army Soldiers from Charlie Company, 1st Brigade, 1st Infantry Division plow a path for a convoy during Decisive Action Rotation 14-06 at the National Training Center on April 20, 2014.

Above: U.S. Army Soldiers from Charlie Company, 1st Brigade, 1st Infantry Division plow a path for a convoy during Decisive Action Rotation 14-06 at the National Training Center on April 20, 2014. (U.S. Army photo by Spc. Randis Monroe/Released)

Lance Cpl. Forest Goodbar, 20, heavy equipment mechanic, Support Company, Combat Logistics Battalion 3, 1st Marine Logistics Group (Forward), detaches a piece of equipment from a M9 Armored Combat Earth-Mover at Camp Dwyer, Afghanistan, Nov. 5. Goodbar and nine other Marines with CLB-3's heavy equipment maintenance section are tasked with maintaining vehicles for several units throughout Afghanistan's Helmand Province.

Soldiers with the 444th Mobility Augmentation Company, New York Army Reserve use their M9 Armored Combat Earthmover to breach an obstacle during a partnered training exercise at Hohenfels Training area during Combined Resolve II.

U.S. Soldiers of Alpha Company, 10th Brigade Engineer Battalion, 1st Armored Brigade Combat Team move their M9 Armored Combat Earthmover behind cover while conducting troop movement training during exercise Combined Resolve IV at the U.S. Army's Joint Multinational Readiness Center in Hohenfels, Germany, May 27, 2015.

Above and below: U.S. Soldiers of 10th Brigade Engineer Battalion, 1st Armored Brigade Combat Team, 3rd Infantry Division construct berms as part of defensive operations training during exercise Combined Resolve IV at the U.S. Army's Joint Multinational Readiness Center in Hohenfels, Germany, May 29, 2015.

Above: A U.S. Soldier of 10th Brigade Engineer Battalion, 1st Armored Brigade Combat Team, 3rd Infantry Division maneuvers a M9 Armored Combat Earthmover to construct a berm as part of defensive operations training during exercise Combined Resolve IV at the U.S. Army's Joint Multinational Readiness Center in Hohenfels, Germany, May 29, 2015.

Spc. John McIntyre, a combat engineer attached to 5th Squadron, 7th Cavalry Regiment, 3rd Infantry Division, stationed at Fort Stewart, Ga., makes a hull-down battle position with an M9 Armored Combat Earthmover during Exercise Peace Sentinel at Novo Selo Training Center, Bulgaria, Nov. 23, 2015. (Photo by Staff Sgt. Steven M. Colvin/Released)

The M9 ACE

On Maneuvers at Ft. Leonard Wood, Missouri

B402

B403

B402

B403

B402
SLING
TIE DOWN
TIE DOWN

B402

SLING
TIE DOWN
TDC 1BDE
35E

TDC 1BDE
35E
B403

TDC 1BDE
35E
B402

B403

TIE DOWN

TIE DOWN

TDC 1BDE
35E
B403

TDC 1BDE
35E
B402

TDC 1BDE
35E
B402

TDC 1BDE
35E
B405

TDC 1BDE
35E
B403

TDC 1BDE
35E
B403

TDC 1BDE
35E
B403

TDC 1BDE
35E
B403

U.S.ARMY
TDC 1BDE
35E

TDC 1BDE
35E
B403

TIE DOWN
TIE DOWN

TDC 1BDE
35E
B402

The M9 ACE

WALKAROUND

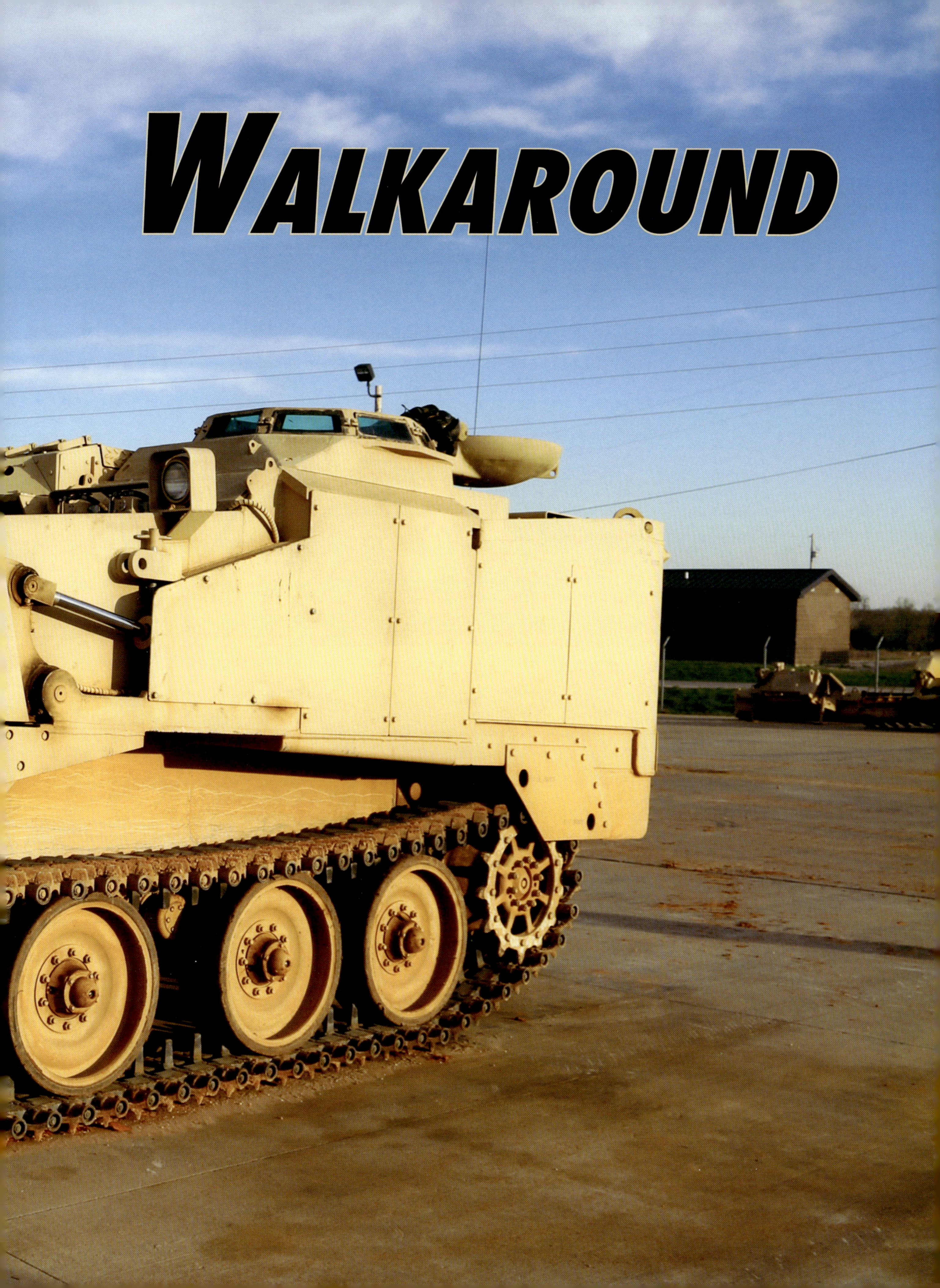

CLAY
U.S. ARMY

B405
17
CLAY

B405

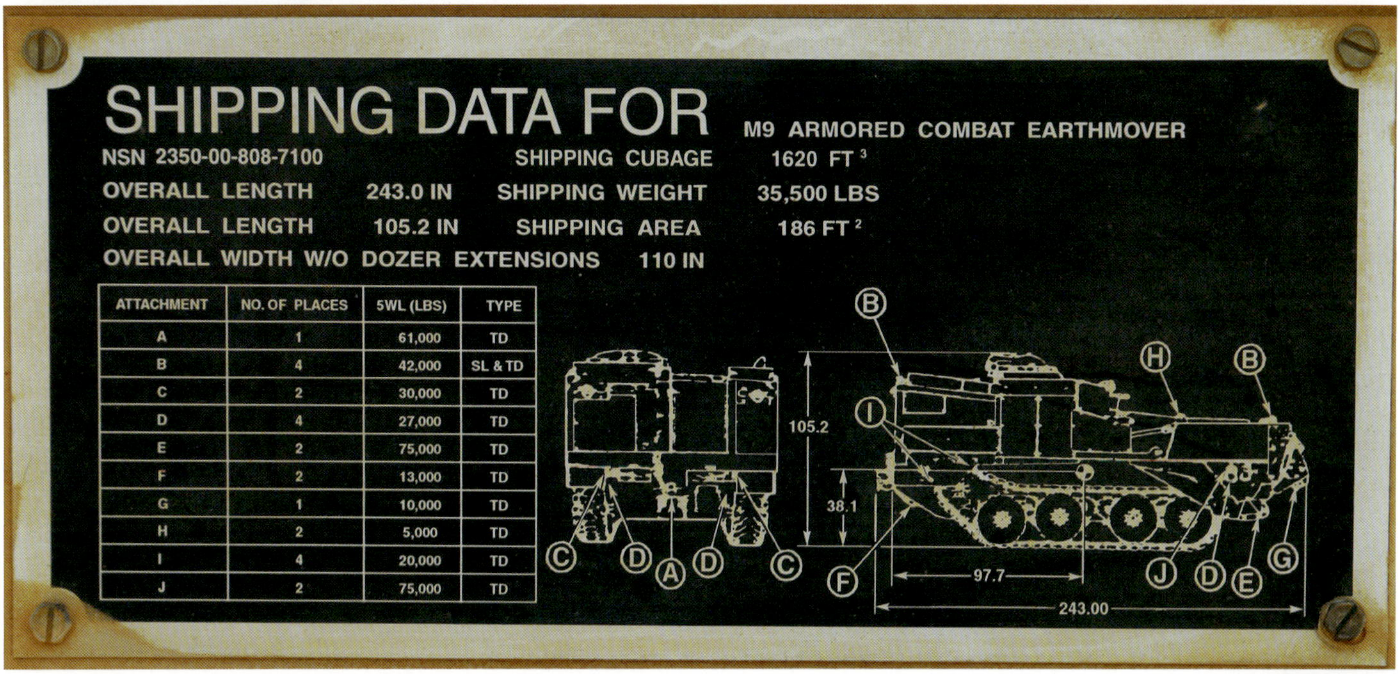

SHIPPING DATA FOR M9 ARMORED COMBAT EARTHMOVER

NSN 2350-00-808-7100 SHIPPING CUBAGE 1620 FT ^{3}

OVERALL LENGTH 243.0 IN SHIPPING WEIGHT 35,500 LBS

OVERALL LENGTH 105.2 IN SHIPPING AREA 186 FT ^{2}

OVERALL WIDTH W/O DOZER EXTENSIONS 110 IN

ATTACHMENT	NO. OF PLACES	5WL (LBS)	TYPE
A	1	61,000	TD
B	4	42,000	SL & TD
C	2	30,000	TD
D	4	27,000	TD
E	2	75,000	TD
F	2	13,000	TD
G	1	10,000	TD
H	2	5,000	TD
I	4	20,000	TD
J	2	75,000	TD

B400

B405

B405

17
B405

SLING
TIE DOWN

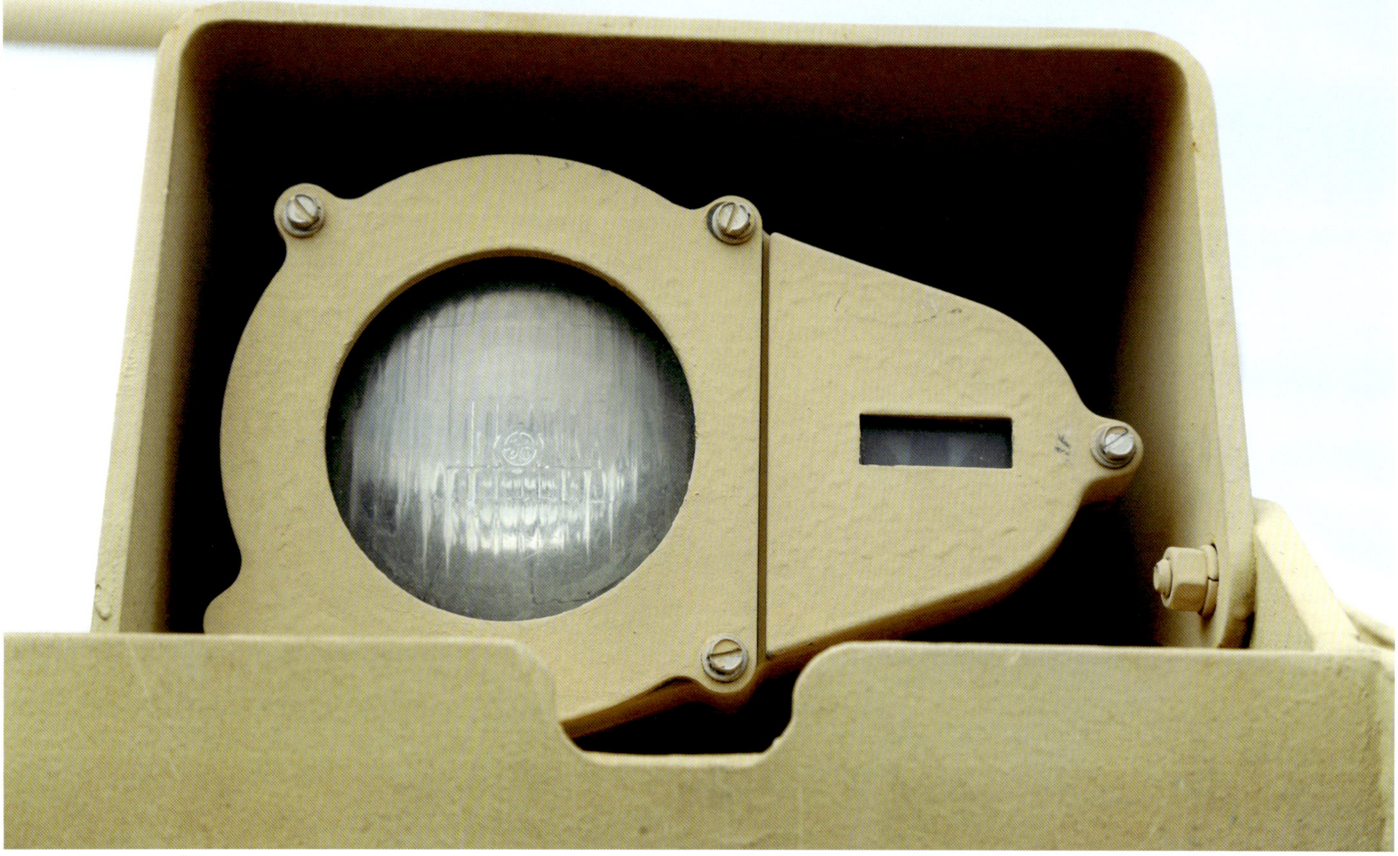

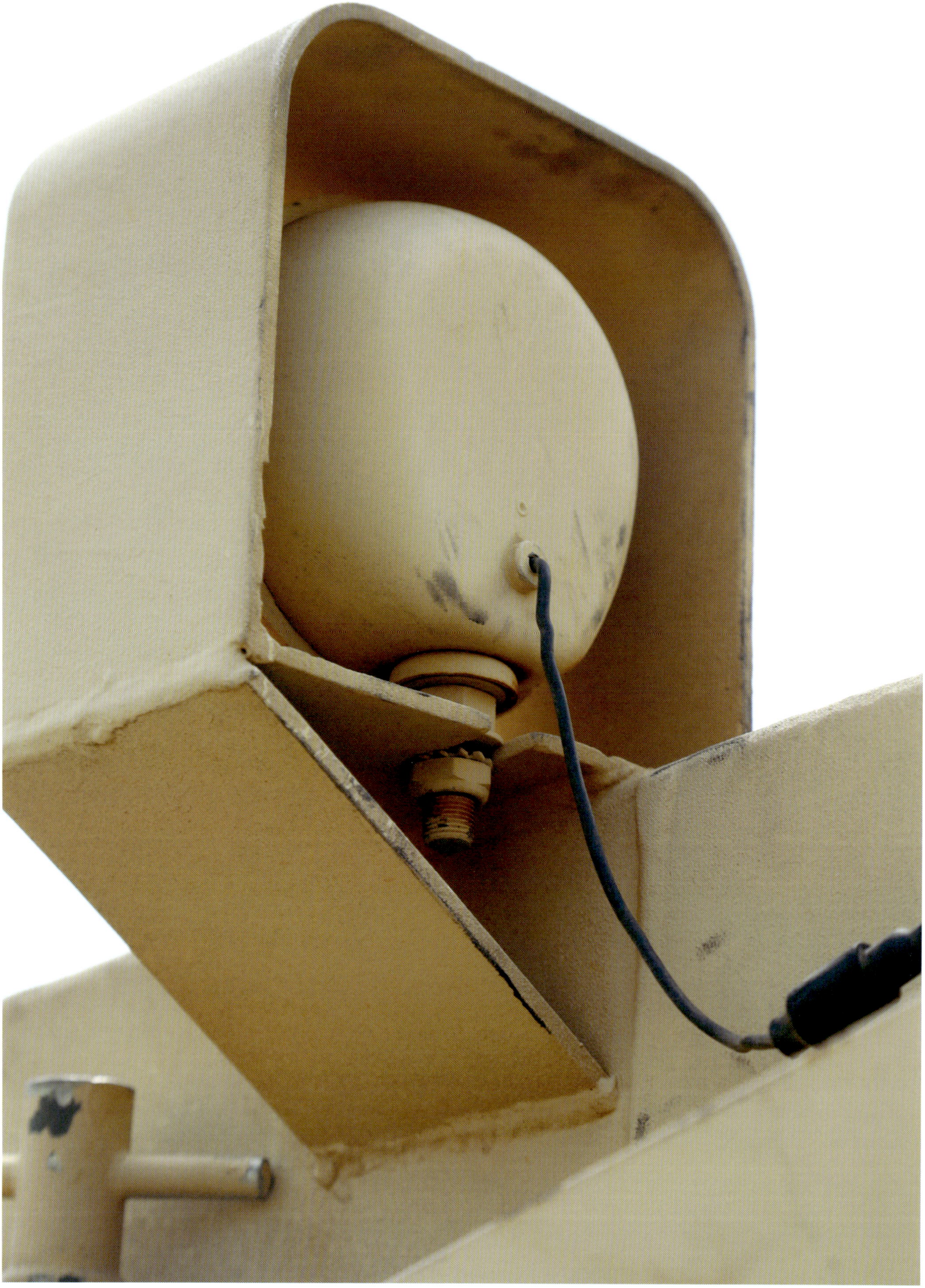

SLING
TIE DOWN
TIE DOWN

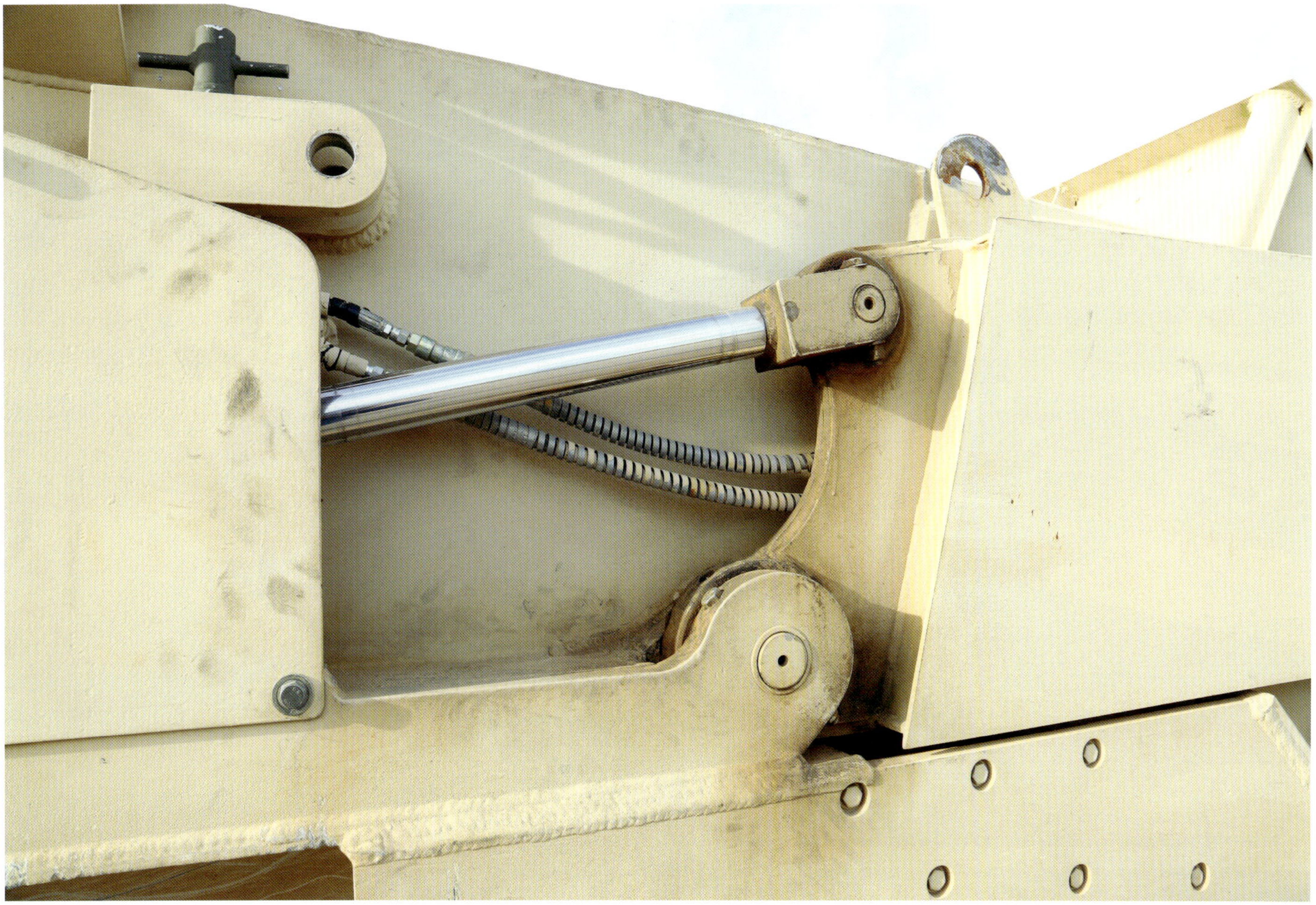

SLING
TIE DOWN
B405

SLING
TIE DOWN
TIE DOWN
TIE DOWN

SLING
TIE DOWN
TIE DOWN

TIE DOWN

SLING
TIE DOWN
TIE DOWN

TIE DOWN

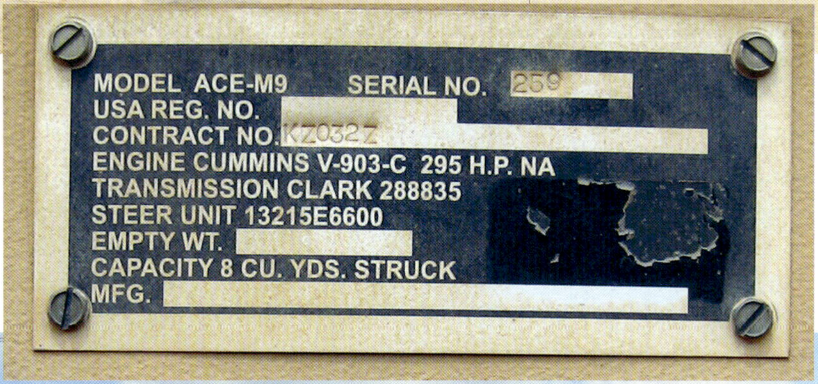
MODEL ACE-M9
SERIAL NO. 259
USA REG. NO.
CONTRACT NO. KZ032Z
ENGINE CUMMINS V-903-C 295 H.P. NA
TRANSMISSION CLARK 288835
STEER UNIT 13215E6600
EMPTY WT.
CAPACITY 8 CU. YDS. STRUCK
MFG.

TDC 1BDE
35E

TDC 1BDE
35E

35E
TIE DOWN
TIE DOWN

U.S. ARMY
KZ037W
TDC 1BDE
35E
B405

U.S. ARMY
KZ037W
TDC 1BDE
35E

EMERGENCY

B405

US
NO STEP

E DOWN
E DOWN

U.S. ARMY
KZ037W
TDC 1BDE
35E
B405

TDC 1BDE
35E

B405
TDC 1BDE
35E
B404

SLING
TIE DOWN
TIE DOWN

SLING
TIE DOWN
TIE DOWN

SLING
TIE DOWN
TIE DOWN

TIE DOWN
TIE DOWN

NO STEP
GOGGLES
N
R1
R2
CB

UNSPRUNG
SPRUNG
MASTER
IGNITION
ENG. OIL
WATER
TEMP
HYD OIL
OUT
BACK
NO STEP
5
4
3
R1
R2
GOGGLES

PERCENT GRADE
GOGGLES
RAISE
MASTER
IGNITION
START

NO STEP
REQUIRED
Wear both
headphones
and earplugs
R1
R2
CB

PERCENT GRADE
GOGGLES
RAISE

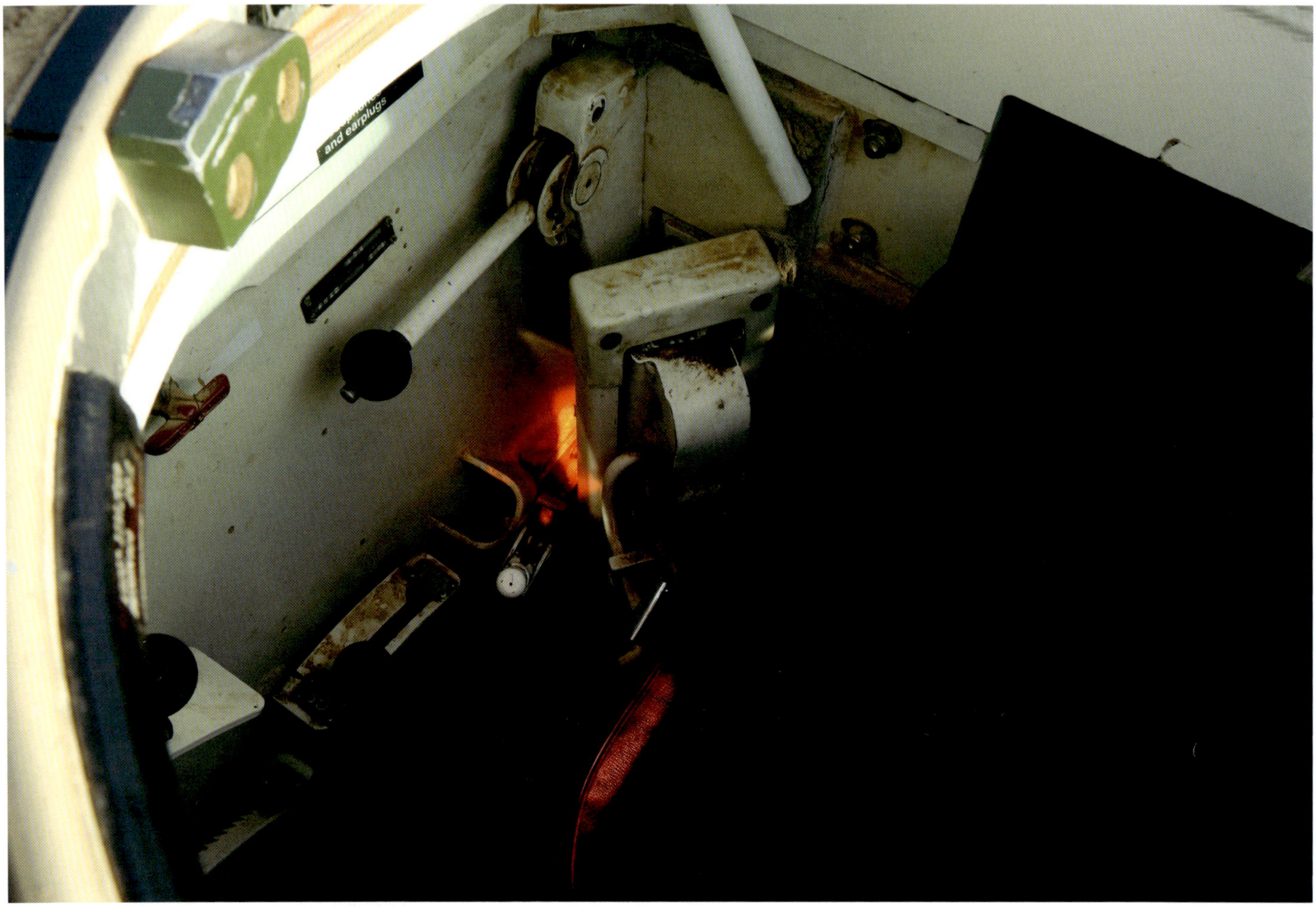
and earplugs

REQUIRED
Wear both
headphones
and earplugs

RIEKER INSTRUMENT CO. PHILA, PA.
PERCENT GRADE
40 30 20 0 20 30 40
EMPTY
1/4 LOAD
FULL
LOAD INSIDE BOWL
R.P.M.
X100
PRESS
WATER

WIPER
START AID
IN
WINCH
OUT
FWD
EJECTOR
BACK
UP
APRON
DOWN
DOME LIGHT
DIMMER
ON

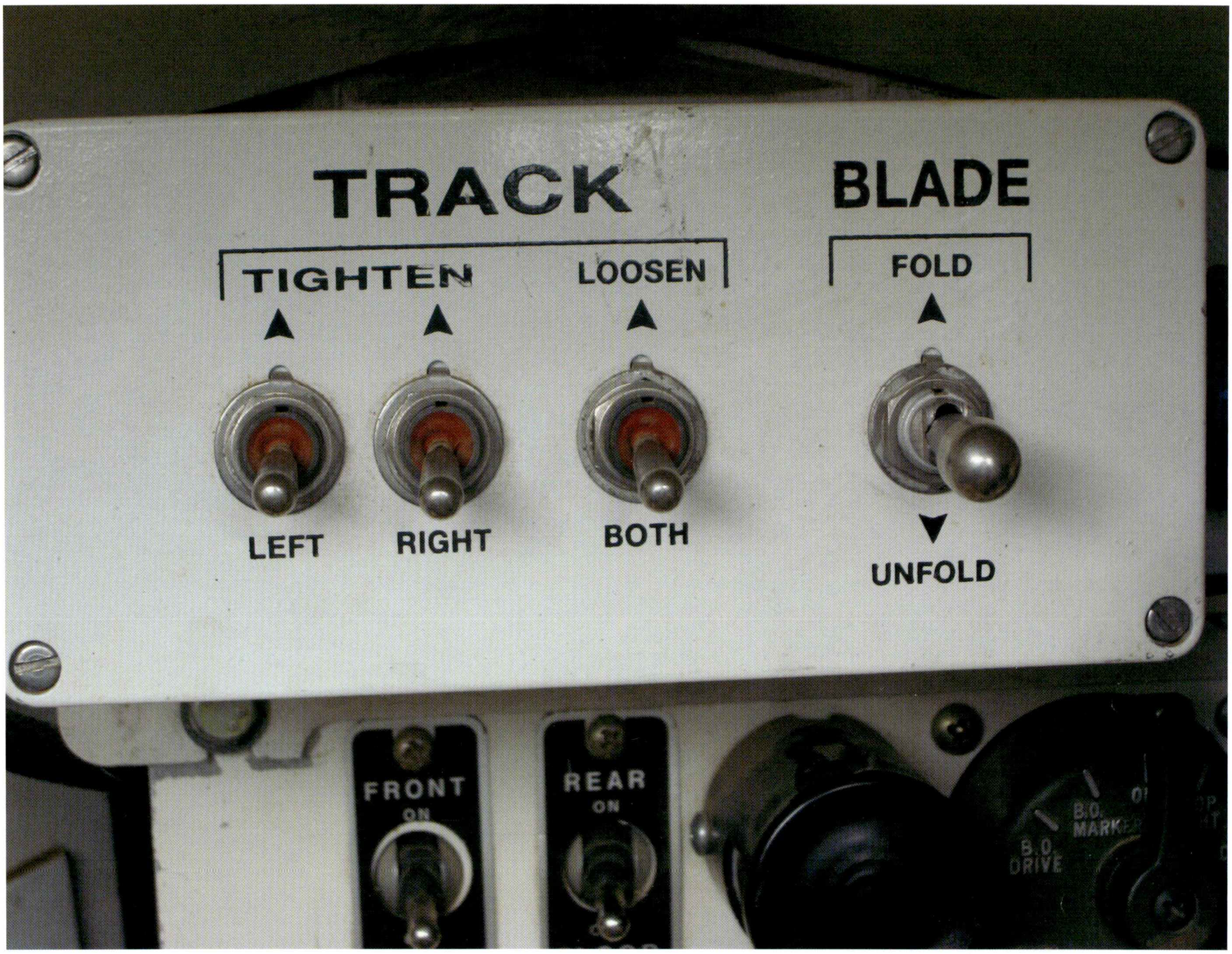
TRACK
BLADE
TIGHTEN
LOOSEN
FOLD
LEFT
RIGHT
BOTH
UNFOLD
FRONT
ON
REAR
ON
B.O. MARKER
B.O. DRIVE

UNSPRUNG
SPRUNG
R.P.M.
START AID
APRON
NO STEP

TRACK
BLADE
TIGHTEN
LOOSEN
FOLD
LEFT
RIGHT
BOTH
UNFOLD
FRONT
FLOOD
REAR
FLOOD
FAN
HIGH
LOW
MASTER
SWITCH
LOW XMSN PRESS
LOW AIR
ON
OFF
IGNITION
START
VOLTS
LEVEL
PARKING
BRAKE
R.P.M.
X100
ENG OIL
TEMP
WATER
TEMP
HYD OIL
WARNING

FLASHLIGHT
WARNING
5
4
3
1
R2

headphones
and earplugs
SERIAL NO.
DATE
OVERHAULED AT

NO STEP

DETEC
KIT

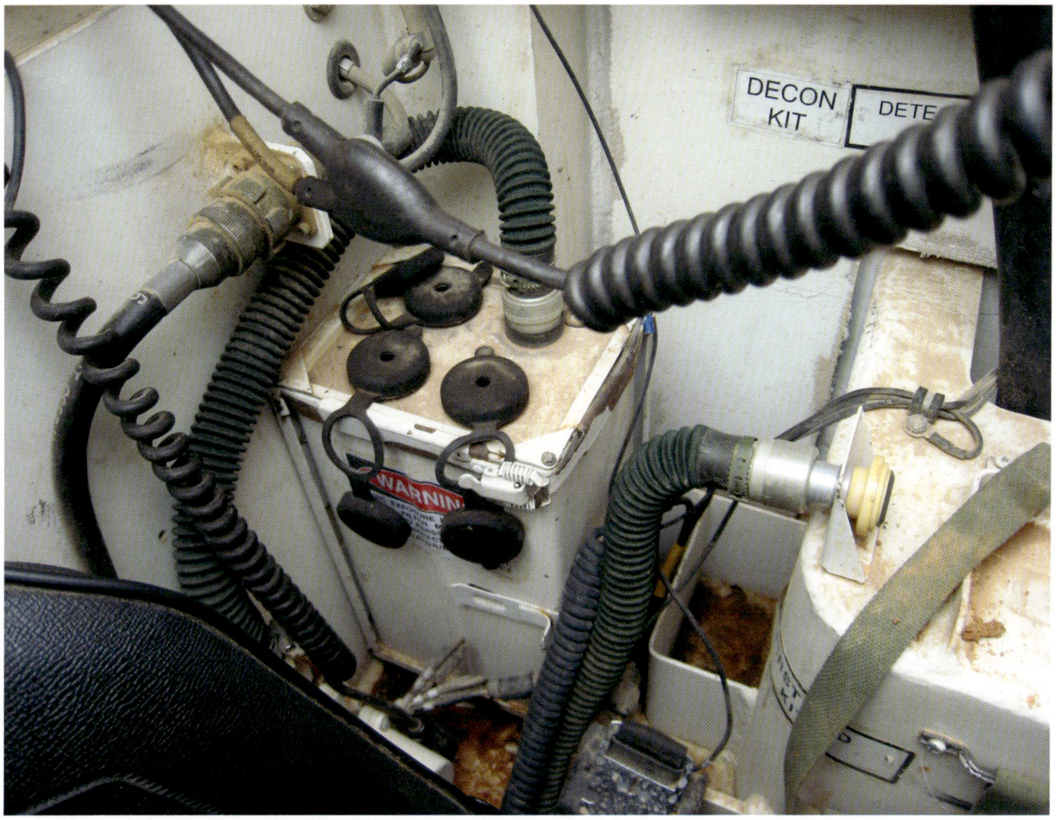
DECON
KIT
DETE
WARNING

J1
ANT
J2
SKILCRAFT
3-RING VIEW BINDER
With clear overlay

RIEKER INSTRUMENT CO. PHILA. PA.
PERCENT GRADE
60
60
30
20
0
20
30
EMPTY
1/4 LOAD
FULL
LOAD INSIDE BOWL

UNSPRUNG
SPRUNG
R.P.M.
X100
HOURS
ENG OIL
WATER
TEMP°F
HYD OIL
MPH
km/h
MILES
UNLOCK
LOW AIR

START
PARKING BRAKE

OFF
1
2
3
4
5
6
7
8
9
FULL
DECON KIT
DETECTOR KIT

WARNING
EXPOSURE IS

ENGINE COOLANT ONLY
DECON APPARATUS

V21
MAIN
PUMP
V22
MAIN
PUMP

V21
MAIN
PUMP
V22
MAIN
PUMP

V21
MAIN
PUMP
WARNING

V21
MAIN
PUMP
V22
MAIN
PUMP
WARNING

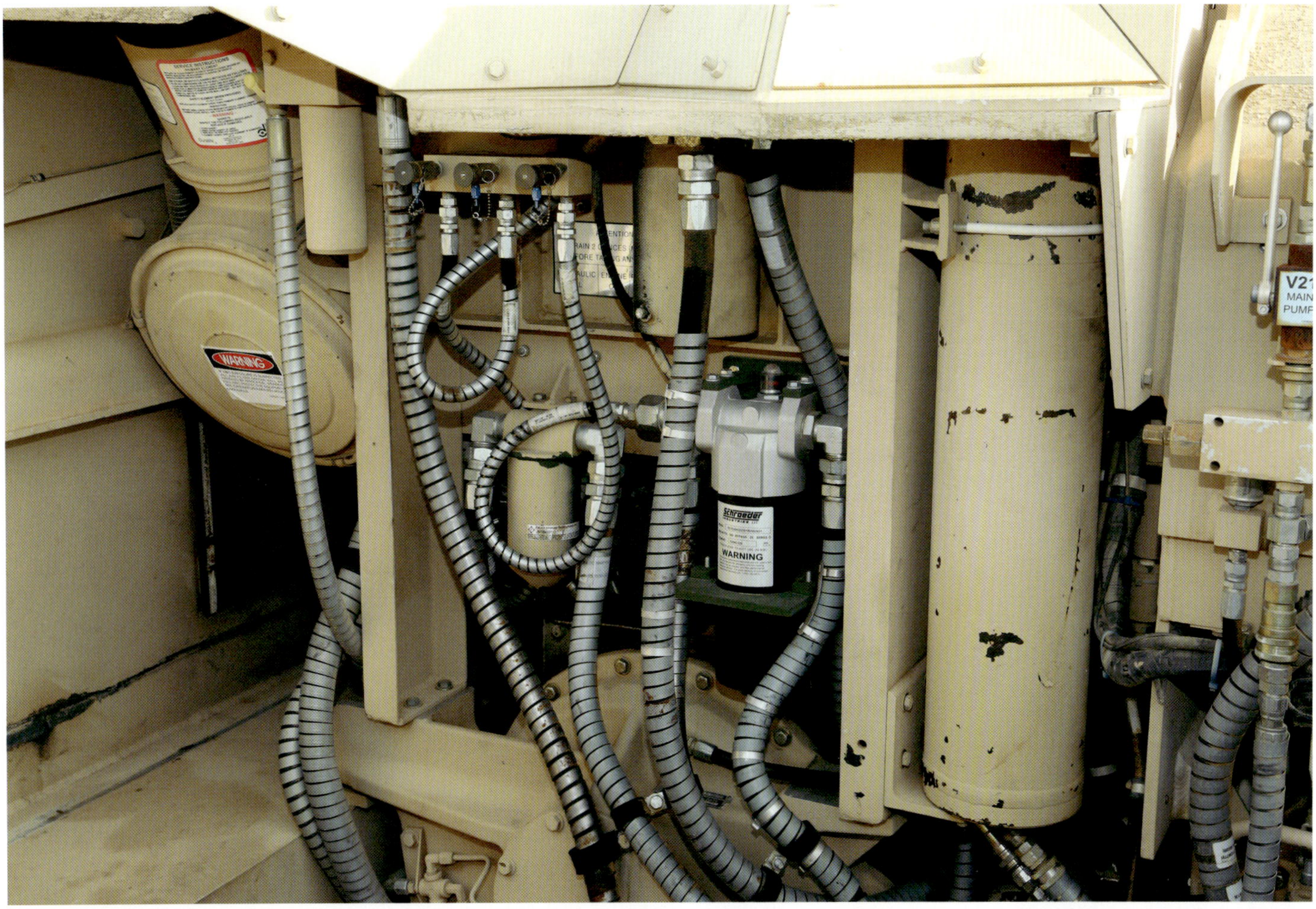
WARNING
WARNING

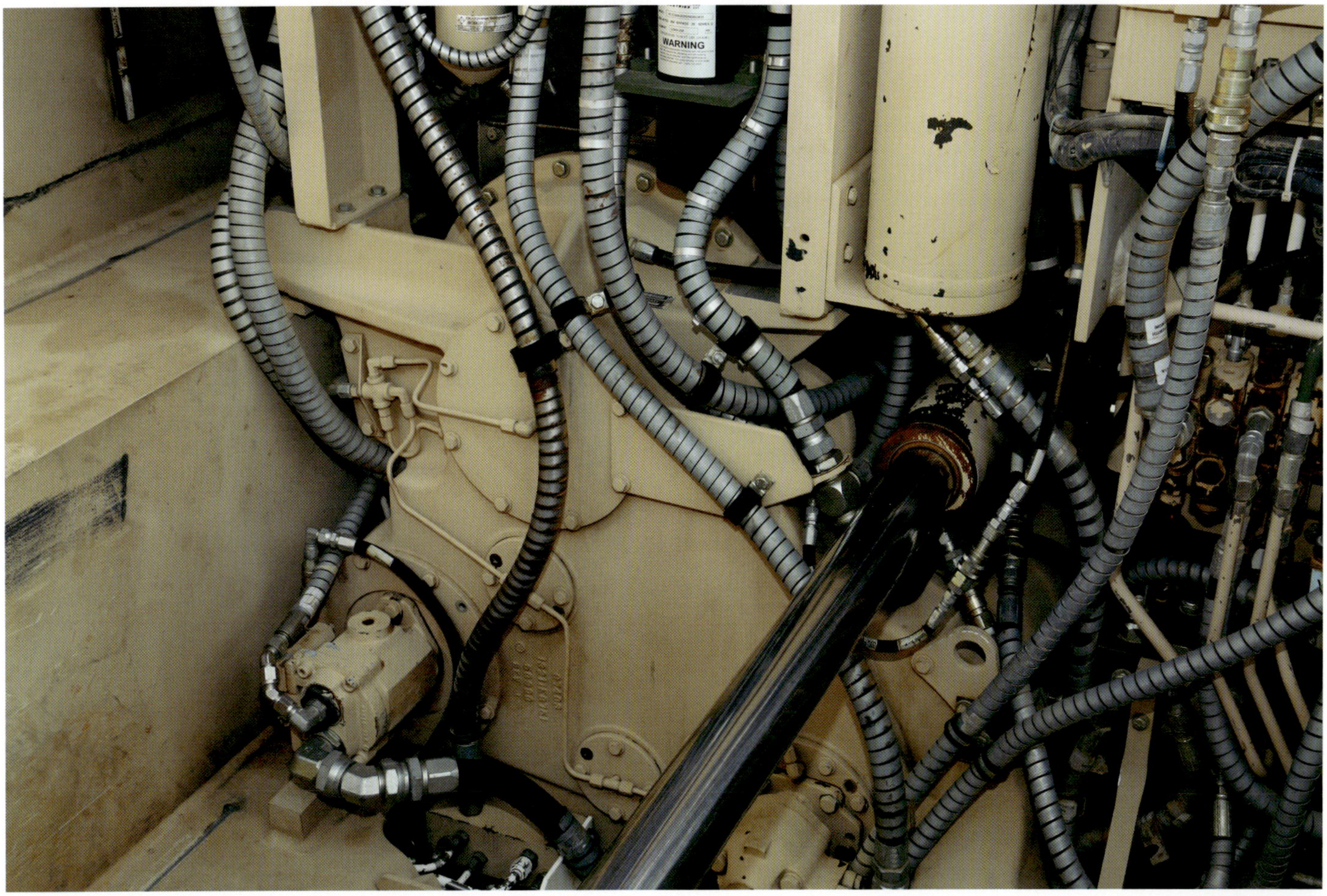
WARNING

V21
MAIN
PUMP
V22
MAIN
PUMP
WARNING

SERVICE INSTRUCTIONS
WARNING
WARNING
DRAIN 2 OU
BEFORE TAK
YDRAULIC
OIL

WATER CAN
DECON APPARATUS
ENGINE

US
WATER CAN
DECON APPARATUS
ENGINE COOLANT ONLY

M9 ACE on maneuvers at Ft. Leonard Wood, Missouri, USA, in April 2016. Note the red clay and dirt, where the earth appears on the vehicle that has been used for a few hours.

SLING
TIE DOWN

TDC 1BDE
35E
B402

Coming Soon!

M1 ABV

Assault Breacher Vehicle

In Detail